Thinking, Periodically

Poetic Life Notions in Brownian Motion

Mala L. Radhakrishnan

Illustrated by Mary O'Reilly

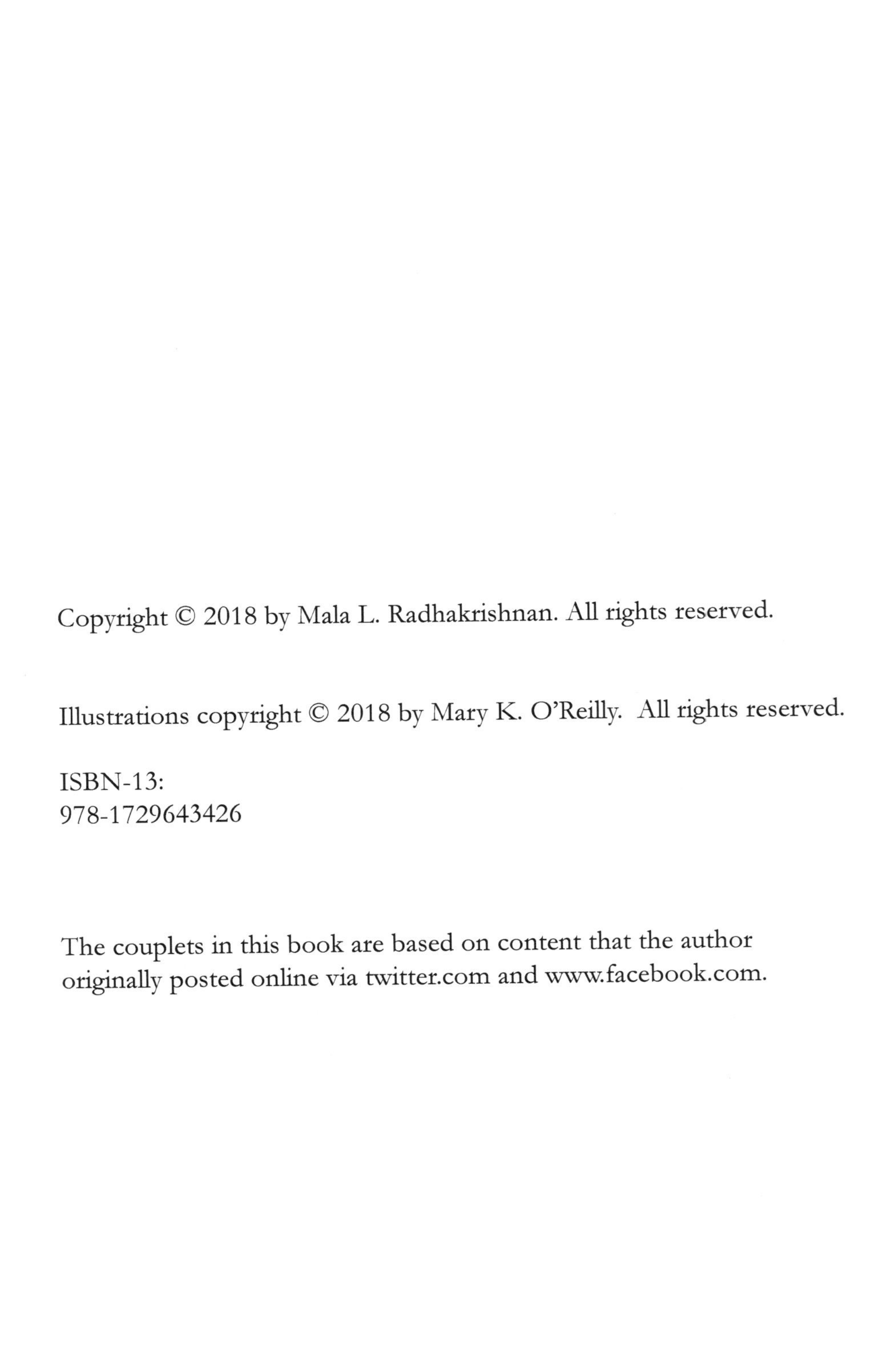

ISBN-13:
978-1729643426

The couplets in this book are based on content that the author originally posted online via twitter.com and www.facebook.com.

For Anjali

Like a magical photon so boldly behaves,
Be bright, be quick, excite, make waves!

Acknowledgments

Through ideas, critique, analysis,
They helped us with catalysis:

Dave Lahr, Lucy Liu, Liz Parker, Helena Qi, Hari Sreenivasan, and the Wellesley College Community

With love that's greater than any can fathom,
Dwarfing the bonds between nitrogen atoms:

Mom, Dad, Anand, Sohil, Samay, and Anjali (M.L.R.)
Mom, Dad, Kevin, Kyle, Lucas, and Sam (M.O.)

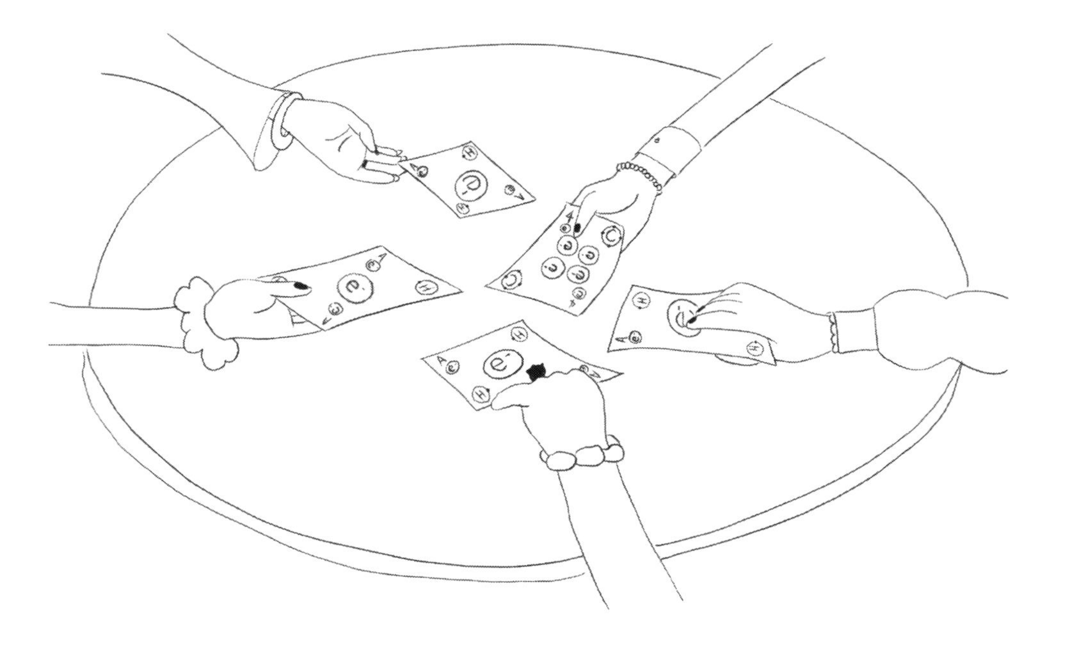

Life is like chem, 'cause the goal's to be able
To bond well with others so we can be stable.

Having control is a major illusion;
We're all at the mercy of random diffusion . . .

Gases and junk have in common a lot:
Both expand to fill up all the space that they've got!

Constructive critique: it reminds me of light.
It only excites if the frequency's right.

Life's like an uncalibrated pipette:
You really don't know just how much you will get . . .

Chemists: they bring to the world greater merriment,
Cycling through model, prediction, experiment.

Creativity's much like a gas in its measure:
It shrinks when its system is under great pressure.

Alcohol: cat'lyst for many events
Whose energy barrier is . . . Common Sense!

Life is like chem lab, I hereby confess;
The reaction you get can be far from your guess.

Kids are like hydrogen—small and light,
And they swiftly collide, react, and ignite.

When a product proclaims that it's "chemical-free,"
Then it's just empty space! Don't you agree?

Teeny, excited, emitting a lot:
A baby's a bit like a quantum dot.

"Two roads diverged in a wood," Frost wrote;
Were I an electron, then I would take both!

A positive force that makes everything better,
A mother's the nucleus that holds us together.

Write books with four letters? 'Twould bring authors strife,
But DNA's four letters narrate all life.

Life with no sleep and no breaks and no naps
Gives new meaning to causing "wave function collapse."

~~~

In quantum we prove we can't know what comes next;
Then in stat mech we prove that we do! Perplexed?

~~~

A walk helps one think so much better, it's scary;
It's brain activation through allostery!

Few acts have an energy barrier so steep
As the process of getting a toddler to sleep.

Chemists, we garner attention and traction
Because we all love to "stir up" a reaction!

Social media followers are swell;
They're like your own personal solvation shell.

My socks are like photons, and I'm not a liar;
They're absorbed and emitted from walls of my dryer.

"I'd love to bond with you, my dear;
Our ψ's constructively interfere!"

How some people juggle their lives leaves me baffled;
They've so many functional groups on their scaffold!

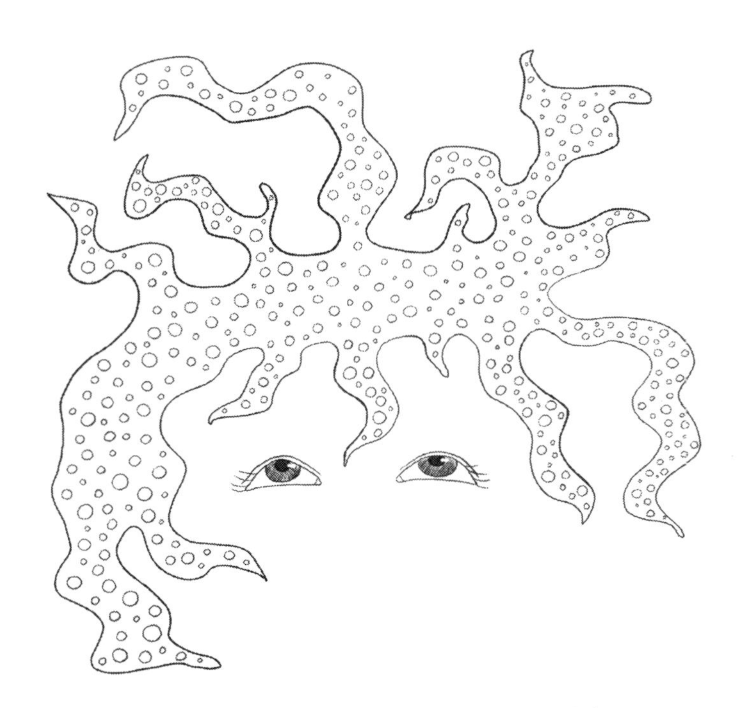

The Second Law? To prove it, I'd say,
"Just look at my hair at the end of the day."

Teachers as catalysts? Somewhat misleading.
Forever they're changed after each classroom meeting.

Stirring reactions removes all the waiting
By giving reactants a chance at "speed dating."

When art meets science, it's truly phenomenal.
Let us stop thinking that they are orthogonal.

To energize family, colleagues, or staff,
Release many joules from a radiant laugh!

Do we all adore chem at first sight? I'd say, "No,"
'Cause for some, it appears to organically grow.

Kids are like 'lectrons—essential yet small,
And I simply don't get their behavior at all.

Extroverts can be uninhibited,
Their interactions diffusion-limited.

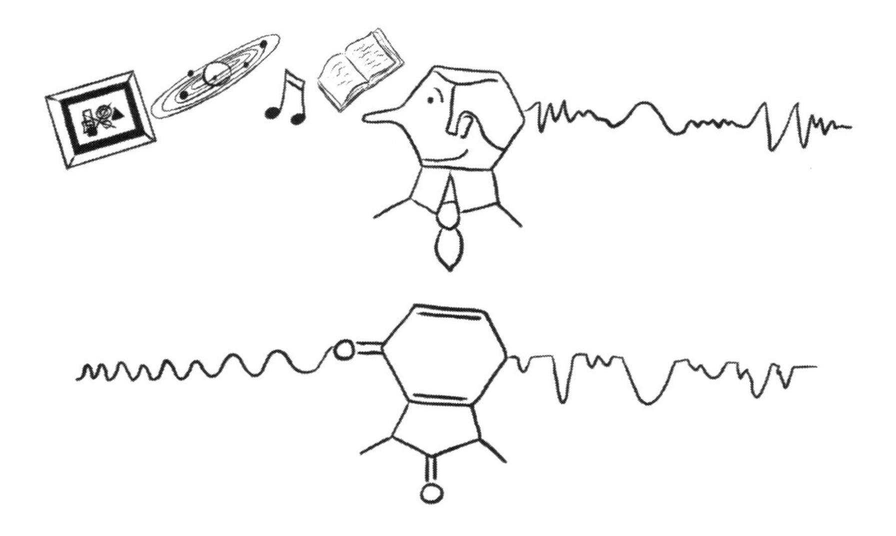

Life's the best match for spectrosc'py by far;
By seein' what excites us, we learn who we are.

Sometimes a compromise leads to stagnation,
A lot like an acid-base neutralization.

Chemistry's drama in which we all revel,
A soap opera on the molecular level!

With some children's eating, we must cut our losses.
Their dinner consumption's a first-order process!

We often forget with our lives on the run
To take time to relax down to n=1.

Life is like research: we strive for perfection
But fizzle and make it our "future work" section . . .

From chemistry do we derive so much pleasure
That even a mole cannot capture its measure!

Showing my stress response? Hooke's Law's a start,
But pull me too far and I might fall apart!

The chance that your days veer off track, are erratic,
As functions of number of kids is quadratic!

Despite what you've heard about Delta G,
Energy's never actually free.

When boiling some water, please do have a heart
And acknowledge the waters you're forcing apart.

In physical chemistry, why, oh why,
Are all of our jokes always met with a "ψ"?

When it comes to life's tasks, please give us direction
On how to achieve catalytic perfection.

Shakespeare on spectroscopy:
"The question: 2B or not 2B?"

If you're not as productive as you'd like to be,
Then you're quite like a kinase without ATP.

Fireworks: brilliant treats for the eye,
With physical chemistry painting the sky!

More truthful than any old chemist's equation:
Some food plus TV yields procrastination.

~~~

Excited are people by sunnier skies!
(It's kind of like they photosynthesize.)

~~~

We chemists imagine that all of our vials
Contain moles of teeny molecular smiles!

You could tell me that it's just the way that they scatter,
But fall leaves break conservation of matter!

Treat others with kindness, respect, and affection,
And you will increase your reaction cross section.

Whilst they are still young, hug your kids every day
'Cause soon entropy spurs their diffusion away.

Success in your lifetime is not a state function
But, rather, depends on your path at each junction.

Chemistry, poetry, life: their relation?
They're all just constrained optimization.

People and atoms are two of a kind:
By their interactions are they both defined.

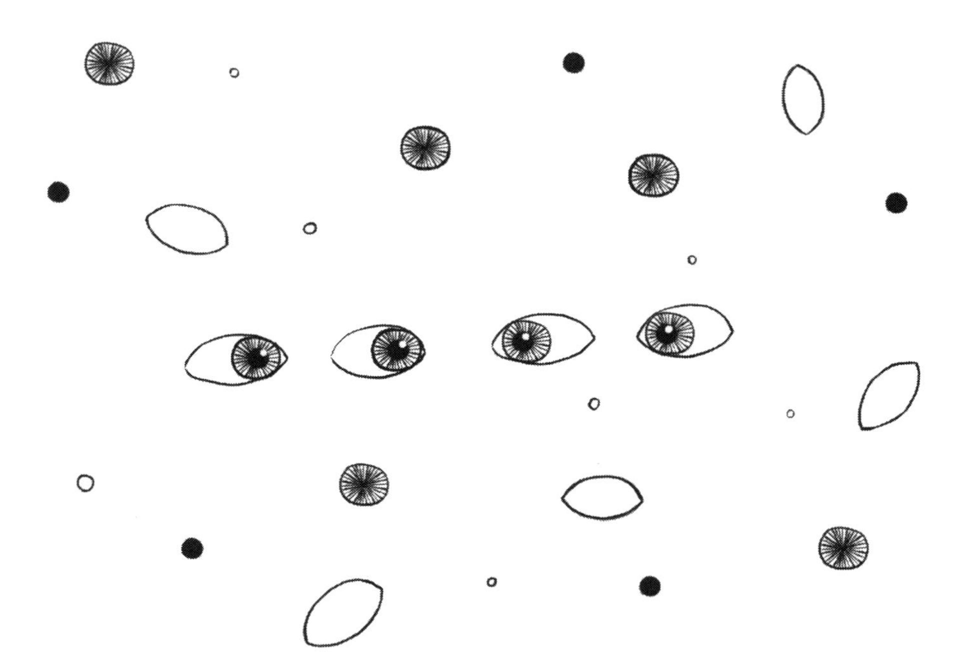

Love at first sight is a potent allusion
To processes limited but by diffusion.

Wanna see entropy's strength in the making?
Show toddlers a leaf pile you spent all day raking!

Babies learn movement through exploration:
Vibration, rotation, and then . . . translation!

Said the rapidly forming distillate,
"Quiet . . . I'm trying to concentrate!"

Classical, quantum—both give me a chill
Because either implies that I have no free will . . .

When sick, there's a chemistry law we ignore;
As our temperature rises, we move a lot slower.

Oh, beautiful particle stuck in a box,
You show us how physical chemistry rocks!

Stay true to your love, if a chemist may say so:
A stoichiometric 1-to-1 ratio.

Though we often pretend that our plans are invincible,
Our lives have a built-in uncertainty principle.

Parents and synthetic chemists equated?
They both can be proud of what they have created!

With kids, "highs are higher, and lows, they are lower."
The thermo is great, but each task is much slower!

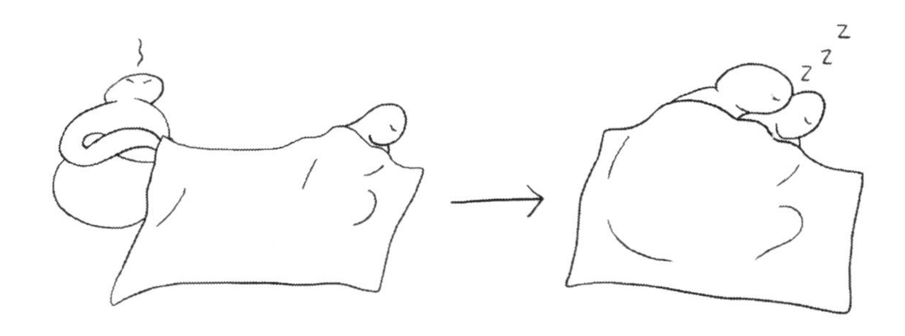

A marriage requires adapting a bit;
It's not lock and key but instead induced fit.

"Just keep moving forward," the optimists say;
For chemists, it's "Just keep your Q less than K!"

Excitement fades fast from material presents:
Fluorescence—or at best, like phosphorescence.

My son loves his noodles, but rather than eat 'em,
He probes their vibrational degrees of freedom.

"Organic" means "natural", "safe" to the masses.
To chemists? Instead, it's gloves, coats, and glasses!

Le Chatelier's principle eludes me, I guess,
'Cause I don't bounce back very well from stress!

When folks act like fluorine, then this we agree on:
It's best to stay calm and unruffled, like neon.

One's tasks and chemistry bear some relation:
They're faster with increased concentration.

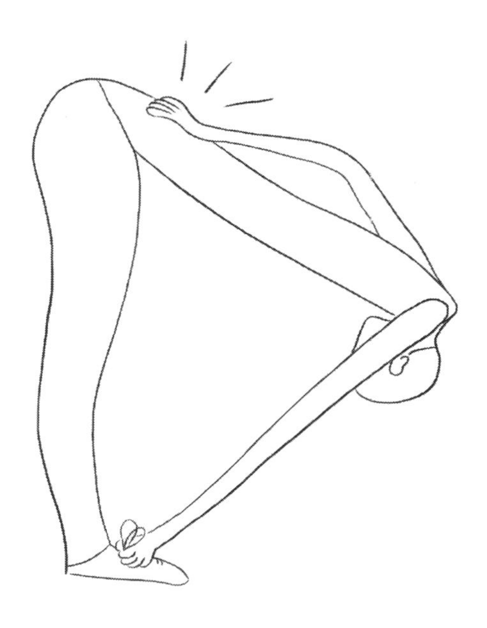

As we age, we become more like cyclopropane,
Forever in some conformational strain . . .

Parents, protecting groups—similar missions:
Shielding their close ones from harsher conditions.

The partition function, a charge, or some heat?
The multiple meanings of "q" all compete.

Embrace Quantum Mech and you'll be off the hook
'Cause your house is a mess only if you dare look!

I'm bad at kinetics. When guessing "how long?"
I always predict 'twill be fast . . . but I'm wrong.

With your arms, to a friend give a hug really big, and
You'll feel somewhat like a bidentate ligand.

Can't handle more input? Well, what do you know?
You've currently got too high a LUMO.

When your brain can't accept a new contribution,
It's much like a saturated solution . . .

The constraints within chemistry are like those in verse;
They're usually a blessing, occasionally a curse.

Though stable a nuclear fam'ly may be,
Beware of radioactivity!

Egos and alkali metals have ties;
They're unstable, reactive when too big in size!

If right now you're feeling a little too blue,
You should alter your wavelength (and get a new "nu").

Whenever my work-life balance has lapses,
I just hug my son—that wave function collapses!

A sleep-deprived brain's not in working condition;
It's lacking both accuracy and precision.

A winter excuse for procrastination?
"No thermal energy of activation."

Of molecular orbitals often we talk,
Yet they're but a model, of Hartree and Fock.

There's always the teeniest mean free path
'Tween the cups, toys, and ducks in a little kid's bath.

When you're finally in your "vacation mode,"
For your ψ should your cell phone serve as a node!

Your choices to no one should e'er be subordinate
'Cause you should control your reaction coordinate!

If our energy's low but high pressure we feel,
Then, just like a gas, we will not be ideal.

What do people and alkali metals both share?
A love for exposure to fresh, outdoor air!

If Calories all went toward heat, it would seem
That some sixty-five burgers would turn me to steam.

"My wonderful darling, of you I'm so fond.
We've a strong, covalent, triple bond."

The Second Law will indeed have its way:
I'll be low E, high S by the end of the day.

Though the nagging to get there should give me compunction,
Once kids are asleep, it is all a state function!

You'd think that in winter, with temperatures low,
All our lives would get stable and ordered . . . but no.

Diamonds and parenting: it's a close call
On which is the hardest thing of all!

Think about all the close friendships you halt
Every time you dissolve a small crystal of salt . . .

If practicing chemistry heightens your terror,
Remember it's sometimes just trial and error.

Like the solid to liquid transition of water,
My heart simply melts when I hug my sweet daughter.

One's first gray hair is an imposition,
A slow, unavoidable phase transition . . .

The contents of baby's colon and bladder
Can violate conservation of matter!

Perhaps we'd unite and we'd see beyond borders
By modeling people to only first order.

If we were electrons, perhaps we'd be merrier
'Cause maybe, we'd tunnel through every life barrier.

Hope you will find this here pun somewhat relevant:
When you study chem, you are quite "in your element!"

The plan for one's life may appear to have vision
But always falls victim to random collisions.

Calculus—useful to both you and me;
It's an integral language in chemistry!

Take action and be whom the others rely on,
And don't simply be just a spectator ion.

People are not much like gases, I feel,
'Cause only when we interact, we're ideal.

In life do we all have a singular purpose:
To probe our potential energy surface.

Upon graduation, one's future awaits
With newly accessible microstates!

To deadlines and due dates we'd all like to keep,
But the limiting reactant is always our sleep!

The journey of parents—a wonderful mess:
A negative dH and a positive dS.

As youth, we diffuse around, plotting our fate.
When older, some enter the crystalline state.

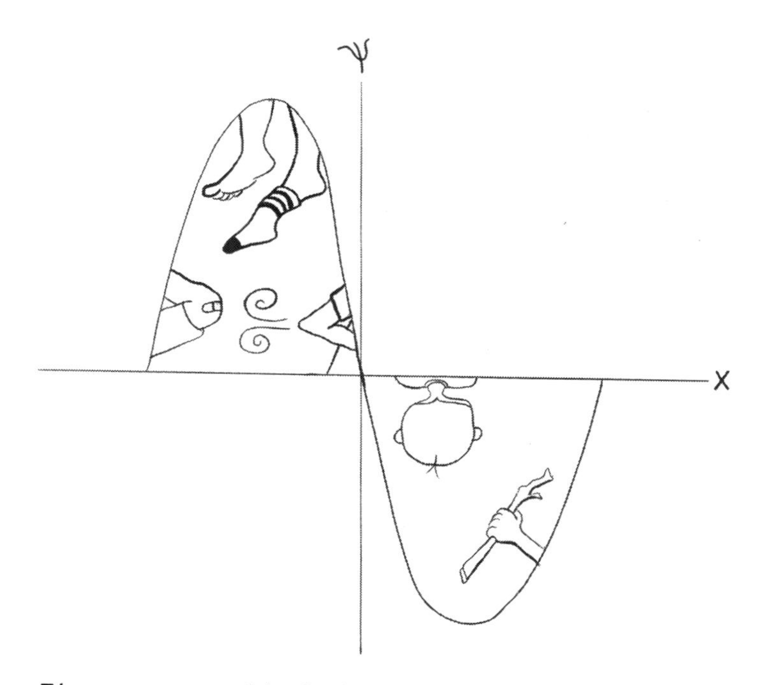

Electrons and little boys bear some relation
'Cause neither one has a defined location.

Cleaning is futile. Denial's a flaw
'Cause there's no use in fighting the Second Law.

When life throws you lemons, right at your face,
Just neutralize them and make courage your base.

Have no direction? Feel lost with confusion?
Just go with the flow—leave it up to diffusion.

On warm days the kids go outdoors with such gumption
And boosted translational partition function.

Oh, hydrogen bonds that our eyes cannot see,
You determine a snowflake's symmetry!

Movies are art, but they have a reliance
On crucial advances in physical science.

"The kids are so quiet now!" Rules adherence?
More likely: destructive interference!

If a flicker of calm you have transiently beckoned,
Your moment of inertia is . . . that very second!

"Indeed, my love, you're so hot, your kT
Makes your maximum lambda lie in the UV!"

All that matters or all that's matter?
While wonderful, chemistry's only the latter.

A chemist creates, and she uses her heart
Because doing good science is really an art.

Iodinated, he boasts of his size:
"My goodness, I cannot believe my own I's!"

Chemically explaining emotions seems tragic,
And yet, the chemistry of love is like magic!

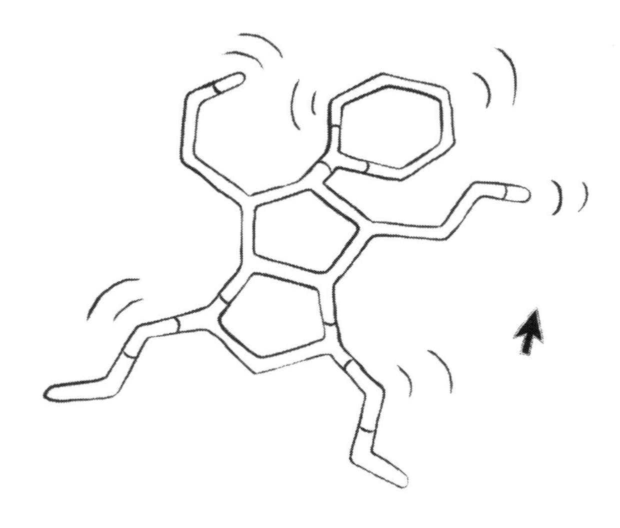

Chemists are artists with every advance,
Choreographing molecular dance.

From metals we learn about happier living
'Cause positive outcomes result from our giving.

You wanna connect more with others? Just try,
And turn up the magnetic stir bar to "high."

Life's like numerical integration:
Too big of a step and you change the equation.

From cheerful and bright to a meltdown condition;
For kids, it's the speediest phase transition.

Your routine's too rooted in repetition?
It's got periodic boundary conditions.

"Twinkle, twinkle little gem,
Thanks to wonders of p-chem."

To catalyze your joy, the key
Is simply doing chemistry!

Accepting life's chaos allows for true livin'.
Just go with the flow and be entropy-driven.

My daughter's loving gaze and grin:
A thermodynamic global min.

The best of the chemists, they know what it takes:
The courage to gracefully make some mistakes.

Mothers (like nuclides): we go through great fuss
To ensure that each daughter's more stable than us!

When winter arrives we might miss all the q,
And at morning and night, we will miss the h-nu.

For any big hurdle that you might be facing,
Oh, may your kT beat the energy spacing!

Why, for young kids, is every last decibel
Thermally always wholly accessible?

After our weekends we're stable. That's why,
On Mondays, the barrier's extra high.

Mechanochemistry caught in the act:
If you stretch us too far, then we start to react!

Like an atom should be ol' disorganized me
'Cause the more things I lose, the more positive I'd be . . .

Money can't make you love something you hate,
But to catalyze finding your passions, it's great!

Were elements letters, and compounds words,
Then chem would be poetry (written for nerds).

What's a state function that's best for my measure?
It's enthalpy. (I'm under constant pressure!)

Life is like chemistry, right in the making,
Where bonds are in flux, always forming and breaking.

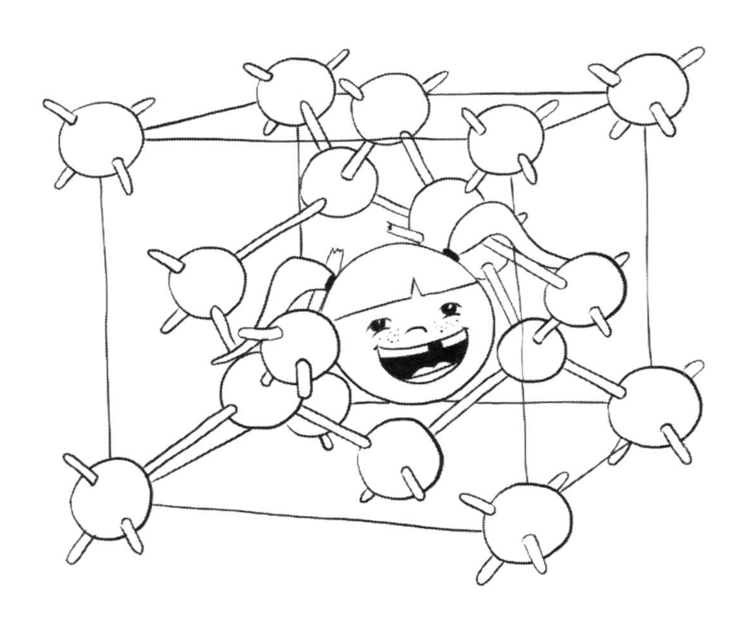

If slight imperfections are getting to you,
Just remember that flaws can give diamonds their hue!

With mol'cules and kids do I play the same role:
Their actions I try, but still fail to control.

Little kids cannot sit still, you see.
They're bopping with zero-point energy.

So easy it is to see chemistry's relevance:
We're all made entirely out of the elements!

When your life has a tough interaction in store,
Be like atoms and don't let it ruffle your core.

A trillion mistakes are not even a few
When you're juggling a mole of tasks to do!

Inherent to data and five-year-old boys
Is that whole unavoidable nuisance of noise.

"Where there's a will, there is always a way."
(Except for kinetic control, I would say!)

Chemistry often has lots of equations,
So don't let your limiting reagent be patience!

An unpredictable car makes you panic.
(Perhaps it was fixed by a quantum mechanic?)

Only a few friends do we know quite well,
While the rest come and go from our valence shell.

Chemists and artists are similar factions.
Both carefully toil, then await a reaction.

When my brain's under pressure, it's saturated.
No pressure? Then it gets evacuated.

A lesson obtained from metallic reactions:
Stay positive always in your interactions.

I behave like an ideal gas by this measure:
I get rather dense under very high pressure.

The events in our life we consider fantastic
Aren't planned or determined; they're often stochastic.

Our faults get passed on to each new generation—
Depressing, like error propagation.

Adults like stability, calm—it's a given.
Children, however, are entropy-driven.

Winter is putting your gloves, hat, and coat on,
At 5PM seeing the sun's final photon . . .

Remain open-minded or life will bring tedium,
Ideas like light on a transparent medium.

If your chemistry passion you need to renew,
Simply have lots of fun with some liquid N_2.

In my growing to-do list I don't make a dent;
My percent yield's perpetually zero percent!

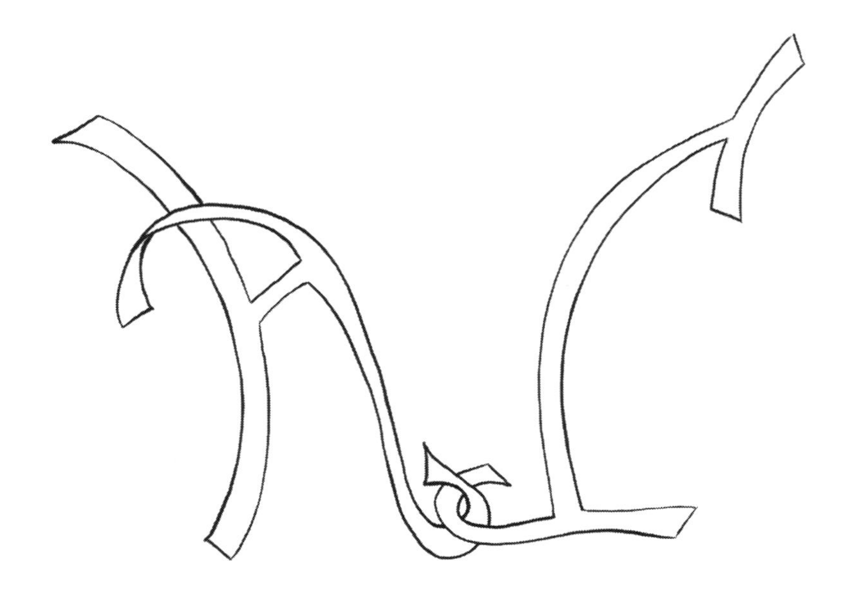

Hydrogen, iodine—both were quite shy,
But they still had the courage to muster a "HI."

"Friends or dating?" You answer with hesitance,
"Complicated—best modeled by resonance."

Look beyond what's expected; develop your vision,
And do not impose any boundary conditions!

Quantum and parenting have me perplexed;
For both, I've no clue what is happening next . . .

Like molecules, most of our time we spend faltering,
But when we surmount a big barrier? Life-altering!

Still want perfection? New parents, stop pining.
Your zero of energy needs redefining.

Do savor the present; if you get too greedy, it
Passes you by like a quick intermediate.

Take baby from Mommy? Such cryin' eyes!
They're an atom that no one can ionize.

The reactant in chemistry that is most relevant
Is not what we think—it's the human element.

Be a poet or chemist, and don't feel inhibited:
Word space and chemical space are unlimited!

Stay afloat when it's rough and few things may seem bleaker,
And never precipitate out in life's beaker.

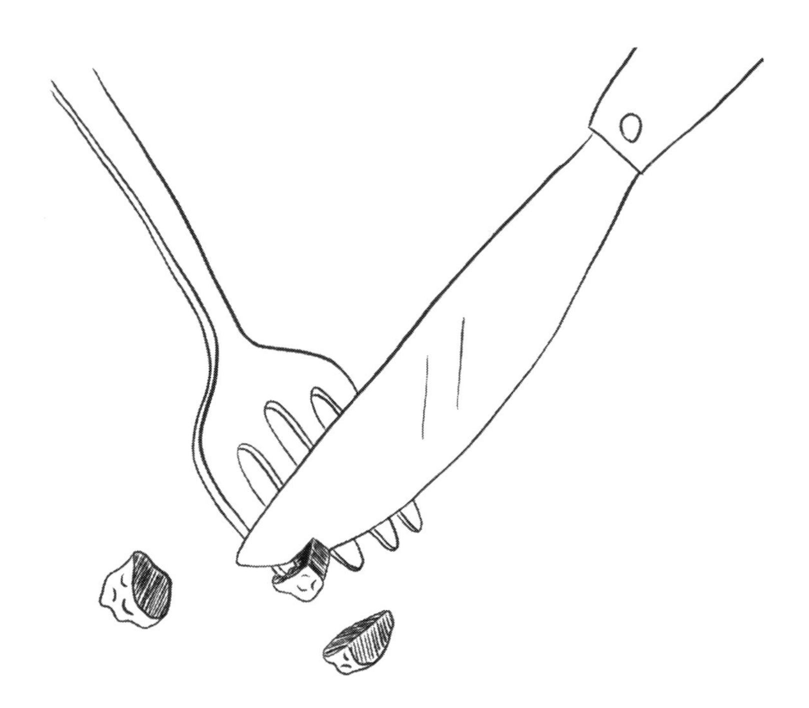

I try being patient with my little dears,
But their half-life for finishing dinner is years.

Solving Schrödinger: a natural high
Involving Me, Myself, and ψ.

If atoms are notes and molecules phrases,
Then chemistry's music that truly amazes!

A child's free energy's never deficient,
With high activity coefficient!

Ign'rance is bliss 'cause you sit with a grin,
Unaware that you're just in a local min.

Driving at rush hour can fill us with wrath
'Cause there's always a miniscule mean free path.

When it's cold, I just wanna curl up in a ball;
My partition function becomes really small.

A toddler who's stuck indoors, you ask?
Like a very hot gas in a stoppered flask!

Obsession with tablets and phones gives new meaning
To chemists' term of "virtual screening."

Patience is trusting kinetic control
When (in seconds) you feel like you're waiting a mole . . .

Weather prediction is one big "Perhaps."
We might as well wait for wave function collapse . . .

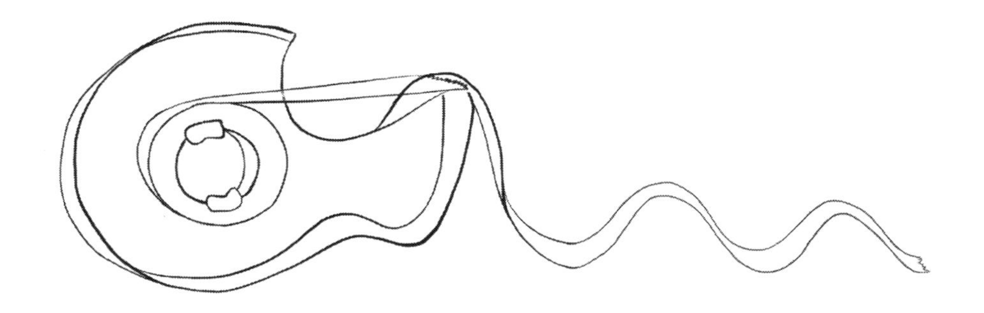

The fastest and easiest thing will take hold.
Kinetics, it has us completely controlled!

A bouncy house with kids I pass,
And nerdy me thinks, "Ideal gas!"

Are chemists abnormal? A refutation:
They always insist upon "normalization."

Life's not black and white; it's the in-between void.
We are each, in a way, like a proud metalloid.

Relax on the weekend, and don't be a hero.
Lay back and enjoy having v=0.

The lower the lows and the higher the highs:
As parents, our ψ's must re-normalize.

"Forever will we remain hand in hand
'Cause you are my complementary strand."

High-pressure explosion you wish to prevent?
I am just like a gas—let me quietly vent!

"You can bring horse to water but can't make him drink."
('Cause a cat'lyst won't help if the thermo doth stink!)

Oscillating 'tween pleasure and strife
Is a normal mode of living one's life.

Toddler's naps? Improbable missions,
Like triplet-to-singlet forbidden transitions.

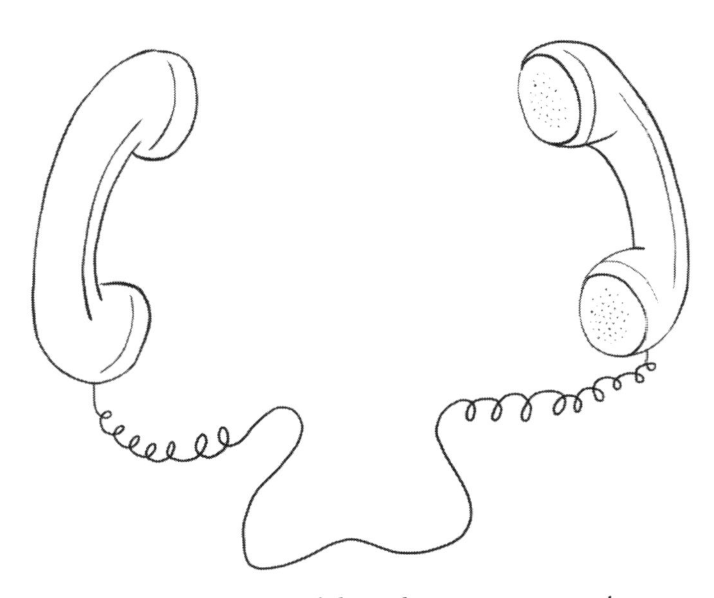

Though your stresses in life often grow to be major,
Don't let them make all of your friendships denature.

Give some kids treats and then watch the commotion.
They can't stop their movements; it's "Brownie-an" motion!

The best demonstration of chem that I've seen
Is my mental acuity after caffeine.

Chemists and parents, they go hand in hand;
Both try to control what they don't understand.

I just can't relax, so am I a fool?
Or do I obey a selection rule?

People and atoms contrast by one measure
'Cause people bond less when they're under high pressure.

If you're overreacting, a chemist might say,
"You should lower your Q so it equals your K!"

We work; we have families and hobbies and mates.
We are superpositions of eigenstates!

When people love chemistry, more is extraneous.
The bonding between them's already spontaneous.

Safety, hygiene, fun, and learning:
Objects of parents' and chemists' yearning.

An organic chemist will readily see
The importance of lots of TLC.

Are Dads merely catalysts? Chemists say, "Never;
They change their kids' thermodynamics forever."

The "last straw?" A chemistry variation:
The ultimate drop of an acid titration.

Do chemists have happiness misunderstood
'Cause they all think that negative energy's good?

"Expressing my love: how close can I get?
I'd need an infinite basis set . . . "

We are like gases—while things seem "ideal,"
The chaos you see right up close is what's real!

Quantum mechanics is simple, you see;
It's as easy as (n equals) 1,2,3!

"How 'bout a date?" Your heart pounds as you wait . . .
The uncertain fate, a transition state!

Though more years to our lives has science been giving,
Humanities make all these years worth living.

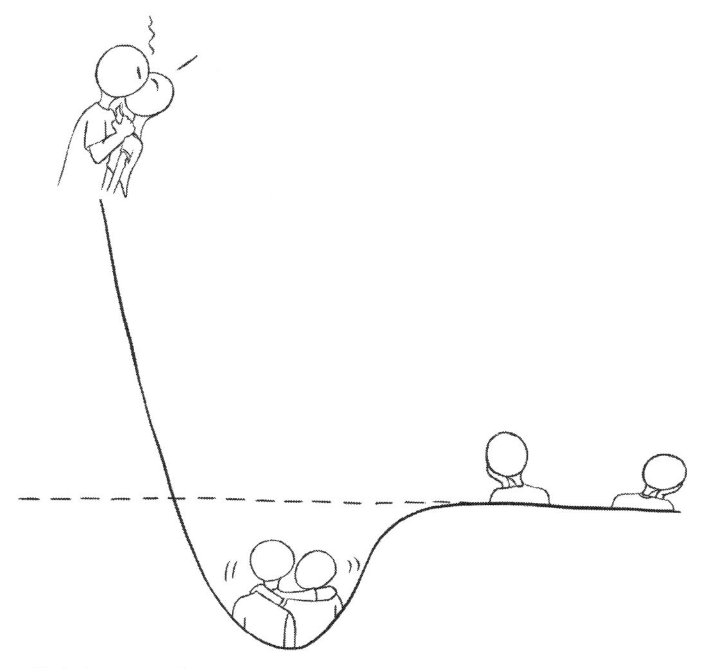

Siblings—like two atoms bonded together:
They squirm and they wriggle, but still, they fare better.

Inclusion and true common ground are elusive
'Cause people are too much like fermions: exclusive.

Life's the chance meetings you have on your path.
(It's collision theory, without all the math.)

A little child's culinary heaven?
$C_{12}H_{22}O_{11}$.

A midlife crisis is when your soul
Feels trapped as if under kinetic control . . .

If you feel somewhat negative, spirits extinguished,
Be not like a 'lectron; be certain, distinguished.

A toddler and fluorine are worth our comparing:
Both little, reactive, and not good at sharing!

Like atoms, our differences help us to thrive
Because multiple elements keep us alive.

For those in our family, we strive for perfection,
Adding our "relative"-istic correction.

Thermo, kinetics—they differ, but how?
It's like "Mr. Right" versus "Mr. Right Now."

Love is a bond, after all's said and done
'Cause two ψ's, they combine to become only one.

Though the path may seem trying to get to your goal,
You should not succumb to kinetic control.

To try out new things must we always aspire
(To make our partition function higher).

While chemistry's great, it cannot explain why
It's the arts that inspire us to laugh and to cry.*

Stretching and yoga help slow down our pace
And sample our conformational space.

Do you remember the thrill and elation
The first time you solved the Schrödinger Equation?

*Yes, sometimes chemistry can make us cry too . . .

Life's like Uranium-238:
If you wanna be stable, you'll just have to wait . . .

Can't seem to relax even though you may try?
Your zero-point energy must be quite high.

Sodium said, "I can't wait 'til I'm able
To find a good mate 'cross the periodic table."

The calm that sets in with vacation's immediate,
But it's just a transient intermediate . . .

When kT is lower outside, we are whiny
Because our partition functions stay tiny.

An isotope substitution? Oh, no!
The victimized, heavier water shouts, "DOH!"

Write "naught equals naught," and although this is true,
A chemist will see it instead as O_2.

It's tough to relax a child for his slumber,
To lower his principal quantum number.

Its complexities thrill us, then make us neurotic;
Our penchant for chemistry's quite periodic!

When proving the Second Law, I am delighted.
Yet still I spend every day trying to fight it!

Technology's like the urea of marriage;
Our chance to connect does it quickly disparage.

Failure should not fill our hearts with such terror.
We're human, and show systematic error!

Life's irreversible—that is the truth.
The Second Law's why we can't get back our youth.

We're just "bags of chemicals"—sounds so pathetic.
To chemists, it's magical . . . rather poetic!

The demands on my energy, I do not need 'em.
I've already too many degrees of freedom!

Social media's like dense noble gases;
We still feel alone as we collide with the masses.

Don't box yourself in; instead, set many precedents.
Be indescribable, like 'lectron resonance!

Some call it "hump day," but chemists, they say,
"It is Wednesday, the energy barrier day!"

It's often quite hard to get toddlers to settle
'Cause they're more reactive than alkali metals.

Lefties: ignored in designs of each fixture
'Cause humans are not a racemic mixture.

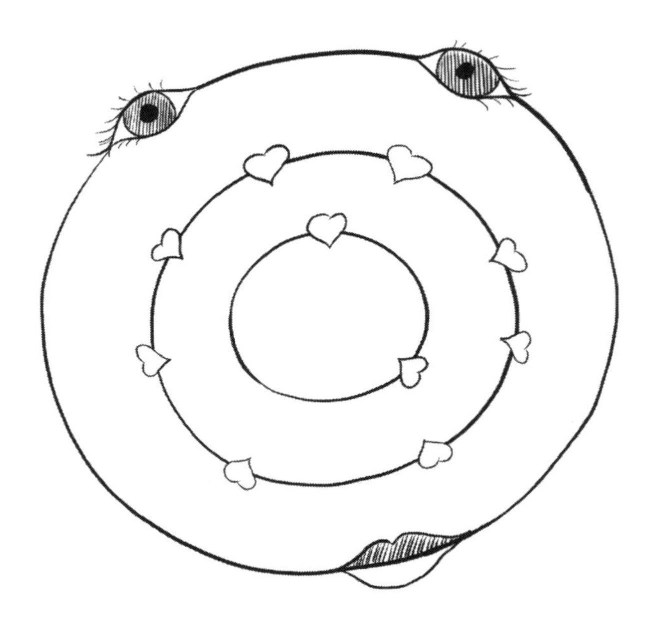

People are shallow—so much we ignore.
We should look beyond valence shells into each core.

Both chemists and parents face tough interactions:
High pressures, big barriers, exploding reactions . . .

"Likes dissolve likes" can describe how we mix
'Cause like molecules, we tend to hang out in cliques.

The love between parents and children, of course,
Dwarfs even the strong nuclear force.

Patience in chemistry should be respected;
The answer will come when you least would expect it.

Find meaning in life despite all the distractions;
Make joy in high yield amidst rival reactions.

Some tasks are quite like a first-order reaction:
Takes eons to finish that last little fraction.

Though the bed is much better, the barrier's too steep
When—relaxed on the couch—you fall quickly asleep . . .

Flailing, reaching, and kicks of elation
Are babies' normal modes of vibration.

Tasks, they keep piling up onto your plate,
Intermediates no more in the steady state.

A sandwich cookie to some is for snacking,
But chemists, they see one and think of "pi stacking."

"Be the change you wish to see . . . "*
e-to-the-x does quite agree.

"Ring Around the Rosie," they sing.
A child and his friends make a six-membered ring!

Physical chemistry: how to get through it?
Believe in yourself and say, "Eigen-do-it!"

*Commonly attributed to Gandhi and Arleen Lorrance

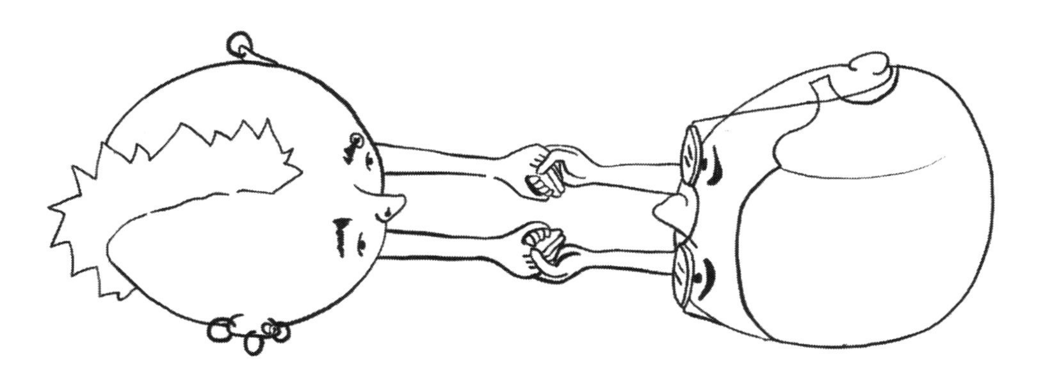

To atoms' relations should ours correspond
'Cause sim'lar or different, they manage to bond.

The entropy's higher for stumbles and fumbles,
And that's why each day of one's life often humbles.

Let's cut across disciplines toward every mission.
We will not collapse as a superposition!

With kids, a new meaning of this I've embraced:
"Hazardous spills of organic waste."

Both molecules, people, on paths they traverse,
Before things can get better, they always get worse!

Chemistry. Poetry. Wonderful things.
Both capture the beauty that every day brings.

Made in the USA
Lexington, KY
22 December 2018

Tew, for being part of all that, and to Christine Calella and Savannah Breckenridge for making sure this title, this cover was on all the shelves, all the lists, in front of all the eyes.

And all the thanks to my wife Nancy, for giving me, what, four months to write this and *Babysitter* and *Mannequins* and *Good Indians*? Four months where, like you always do, you kept my world working, and let me hide in stories. But, I don't always hide there. I also like to live in this story you and I are in. The one where it wasn't just me shooting up 80 to get to our son, but us. The wind was ridiculous, kept wanting to turn us into kites, and none of my radiators were as good as I hoped, and the big rigs were buffeting us across the lanes like playthings, and the ice was, you know, ice, but I had your hand to hold across the console. It should have been on the steering wheel, but you're more important. If we did have to go sailing across the ditch, to cartwheel end over end into Wyoming? I wanted to feel your hand in mine in those last moments. But we lucked through time and again, and each trip, we'd stop at Wagonhound and we'd stand in that bitter-cold wind together, and we'd know that we were in this together, and no matter which way the road turned under us, no matter what big rig was bearing down on us, it would be you and me, Nan.

It still is. It always will be.